Mercury in Dental Fillings

The impact of mercury on health.
Safe removal of dental mercury, and the use of
safe options for restoration of teeth.

An information booklet complied by

Stewart J Wright BDS

Augur Press

MERCURY IN DENTAL FILLINGS
Copyright © Stewart J Wright 2010

The moral right of the author has been asserted

British Library Cataloguing in Publication Data.
A catalogue record for this book is available from the British Library.

ISBN 978-0-9558936-2-9

First published 2010 by
Augur Press
Delf House,
52, Penicuik Road,
Roslin,
Midlothian EH25 9LH
United Kingdom

Printed by Lightning Source

Mercury in Dental Fillings

Disclaimer

This book is not intended to be a source of medical advice. If you are concerned by any of the issues raised, professional medical advice should be sought.

Every effort has been made to ensure the accuracy of the information that appears in this book. However, the author and the publisher do not accept any responsibility for loss or damage arising from reliance on this publication.

Acknowledgements

To all those dedicated researchers, dentists, doctors and patients, whose tireless efforts have brought the mercury issue to the fore and have provided all the information in this booklet.

To all these people, many thanks from all who have benefited in the past and will benefit in the future from this information.

In particular, I want to thank Sam Ziff and Michael Ziff, whose Health Information Booklet 'Dentistry without Mercury' ISBN 0-941011-04-6 is an important asset to all those who have an interest in mercury-free dentistry. I found their book to be a valuable source of references and information during the preparation of 'Mercury in Dental Fillings'.

Foreword

For many years I have been concerned about the damaging effects of mercury from dental amalgam filling, and for this reason I have not used fillings containing mercury since around 1990.

The original reason for discontinuing the use of these fillings was very much a personal one. I had found that my hands and feet were becoming more and more numb, and with some fear and foreboding I went to my GP, who referred me to a consultant neurologist. I thought that I might have multiple sclerosis, and I feared the outcome of the tests arranged by the consultant. After nerve conductivity testing and many blood tests, I was told that I was suffering from idiopathic peripheral neuropathy. I was also told that there was no cure for this condition, and the doctors could give me no reason why the numbness had come about. They advised me not to worry and to carry on with my life. They added that stopping smoking and drinking alcohol might help.

However, I felt that there had to be a specific reason for the problem. I was certain it was something I was doing that might be causing it, and I started to look at the materials I was using at work. At that time I was using a lot of plastic to make mouth splints, and I felt that my condition could have been provoked by handling these materials. I searched the literature, but could find nothing to substantiate this.

Determined to understand what had caused the numbness that I was enduring, I continued my search. Slowly the significance of daily exposure to mercury from working with mercury amalgam fillings came into the picture, and I decided to act on this.

Since then, my life both professionally and personally has changed, and in my quest to help both myself and my patients I have met some wonderful people from many walks of life who have helped me in so many different ways.

While trying to find a way of reducing the amount of mercury in my body, I tried a number of approaches, many of which were of some help. The numbness would recede, but then it would build up again, although not to quite the same level. About three years ago, I decided to try Field Control Therapy (FCT©), and this is the approach that has made the greatest difference to my health. As a result of FCT the numbness is much better, and has gone away nearly completely. A brief description of FCT can be found in this booklet in the section on detoxification.

I can now view the numbness that crept into my life as almost a blessing in disguise. It was the symptom that alerted me to the fact that there was something very wrong, and led me to become aware of the effects of mercury on health. This in turn resulted in my decision to discontinue the use of mercury amalgam fillings in my dental practice, thus benefiting my own health and that of my patients.

Stewart J Wright

Contents

Introduction

The term 'silver filling' is potentially misleading, as such fillings are made from 50% mercury and 35% silver, with the remaining 15% being a mixture of copper, tin and zinc. The mercury is exactly the same substance that has been banned from use in thermometers in all hospitals, schools and throughout industry because of its toxicity.

These fillings were originally used in the UK more than 150 years ago as a cheap alternative to gold, which was the most common material in use at the time. However, most people could not afford to pay for gold fillings, and 'silver fillings' were used as a cheap alternative. Even then there was concern about possible toxic effects of the mercury in the fillings, and various dental associations banned its use at the time, but economic forces won the day.

Apparently, a silver paste containing mercury was used by the Chinese as early as the 7th century.

A 'silver filling' of average size contains approximately 750 - 1000 milligrams of mercury. In addition to the inclusion of silver, copper, zinc and tin are commonly used (1). Consequently these fillings would be more accurately called mercury fillings, as mercury is by far the greatest constituent. 75-80% of all tooth fillings have been of this type (2).

Until very recently, students at dental school were taught that once the filling was mixed, the mercury was 'locked into the filling and would not escape'! The mixing of the mercury with the amalgam metals is conducted immediately prior to inserting the filling.   The British Dental Association has recently admitted that the mercury in the filling can indeed escape.

Escape of mercury from 'silver' fillings

This is caused by the following processes:

- Once covered with saliva, the fillings begin to corrode – a process commonly referred to as 'rusting' – and this releases mercury from the filling as well as silver, copper and tin, all in ionic form, which in themselves are toxic.

- The action of chewing and grinding releases mercury vapour. In addition, microscopically small pieces of filling break off and are swallowed. A high proportion of these minute pieces lodge in the gut, and the metals are slowly absorbed into the body.

- Increases in temperature accelerate the rate at which mercury vapour is released, so hot food and drinks increase the amount of mercury vapour released. Consequently, each time you eat or drink anything hot, you are increasing your exposure to mercury.

- If your mouth has more than one type of metal in it – i.e. gold crowns or fillings as well as 'silver fillings' – then a battery effect can be set up which can generate currents that can be measured in milliamps. There is a risk that an electrical current of such strength can interfere with brain activity. The strong currents produced by activity between the metals in the mouth can also provoke an increase in the release of mercury vapour from the fillings.

Thus 'silver fillings' release mercury vapour into your mouth continuously, allowing immediate absorption of mercury into your body. Some of the mercury vapour may react to produce mercury compounds – inorganic and organic. These are also absorbed, and some of these compounds – particularly those in which the mercury is bound in an organic form – may be even more toxic than mercury vapour itself.

This low level, continuous release of mercury vapour can affect your health adversely, as mercury vapour is very toxic, and even minute amounts are not without effect.

The USEPA standard for safe exposure has been set at 0.1µg per day per kg bodyweight. This means that an adult weighing 70 kg (154 lb) can tolerate only 7µg per day, and a child weighing 77 lb can tolerate 3.5µg per day.

See next page for the estimated average daily intake and retention of mercury and mercury compounds by people who are not occupationally exposed.

The Environmental Health Criteria 101: Methylmercury (WHO 1990)

This document gives the following as estimated daily intakes of elemental mercury (the vapour), inorganic mercury compounds and also methylmercury (a very toxic compound of mercury that can form in the mouth, or can be ingested from contaminated foodstuffs).

See below for the estimated average daily intake and retention (μg/day) of total mercury and mercury compounds in the general population not occupationally exposed to mercury. The figures in brackets are the estimated amount (μg) retained by the body of an adult.

Exposure	Elemental mercury vapour	Inorganic mercury compounds	Methylmercury
Air	0.030 (0.024)	0.002 (0.001)	0.0080 (0.0064)
Food: Fish	0	0.600 (0.042)	2.4 (2.3)
Non-fish	0	3.60 (0.25)	0
Drinking water	0	0.050 (0.0035)	0
Dental amalgams	3.8-21.0 (3.0-17.0)	0	0
Total	**3.9-21.0 (3.1-17.0)**	**4.3 (0.3)**	**2.41 (2.31)**

No government department or authority anywhere in the world has ever certified 'silver fillings' as safe for use in humans.

1996 Health Canada Report. A risk assessment was carried out for the Canadian government by Dr. Richardson (20), a well-respected risk assessor. Basing his recommendation on all the available scientifically validated data, he concluded that up to 2 fillings in children and 4 fillings in adults would be an acceptable risk. He stated that any more than this would be **an unacceptable risk to the health of the patient.**

Bioavailability of the mercury in 'silver' fillings – prevailing dental opinion versus scientific fact

Dental opinion:

Once the mercury is mixed with other metals in the filling it is 'locked in'.

Scientific fact:

Mercury vapour is continuously released from 'silver fillings' (4). An increase in the release of vapour is caused by chewing (3, 4, 5), brushing of teeth (6), and consuming hot, salty or acidic foods or drinks (7).

The more 'silver fillings' you have, the greater your daily exposure to mercury (4).

Once you have stopped chewing, it takes 90 minutes for the rate of mercury release to drop back to its pre-chewing rate (4, 5). Thus you are being exposed to a roller coaster of mercury vapour release all day. Consuming breakfast will cause the release of mercury vapour to rise, and just as it falls, it is time for the mid-morning coffee break, when the release of mercury vapour would be increased again. This pattern goes on all day with lunch, mid-afternoon coffee, dinner and supper.

Dental opinion:

Any mercury from 'silver fillings' is not retained in the body, and is in such small doses that it has no effect on the health of the body.

Scientific fact:

German, Swedish and American investigations (6,8) have found that human brain and kidney tissues from people with 'silver fillings' contained more mercury than those tissues of people without 'silver fillings'. The amount of mercury in the brain tissue of the people with 'silver fillings' correlated with the number of fillings they had – i.e. the more fillings the person had, the higher the concentration of mercury in the brain tissues.

It has now been established that the mercury from dental 'silver fillings' constitutes the largest single source of inorganic mercury exposure in the general population, and is greater than all the other environmental sources of inorganic mercury added together (10). Despite replicated scientific findings, dental authorities maintain that 'silver' dental fillings are safe. They base this on the fact that the fillings have been in use for 150 years and that dentists have not seen people who are sick or who have died as a result of these fillings, and yet they do not make it clear that dentists are neither trained nor licensed to determine mercury toxicity (11).

However, from the medical perspective, 'silver fillings' are considered to be a significant source of mercury, thus having a significant potential for toxic impact. Medical researchers are now investigating the possible health risks of mercury from 'silver fillings'.

The latest medical research shows that mercury from 'silver fillings' accumulates in all the adult tissues, being highest in the kidney and liver (13). Research also demonstrates that mercury crosses the placental membranes and accumulates in the developing baby within two days of having a filling placed (14).

Although there have been reports to suggest a relationship between the presence of 'silver fillings' and a reduction in health of humans, it is only recently that science has established a direct cause-and-effect link between 'silver fillings' and pathology.

Experiments carried out at the Departments of Medicine, Pathology and Physiology at Calgary University by M.J.Vimy, N.D.Boyd, D.E.Hooper and F.L.Lorscheider (21) show a 50% reduction in kidney function in sheep one month after inserting 'silver fillings'. Kidney function was found to continue to fall after 60 days. The kidneys remove harmful substances from the blood, and maintain blood pressure and fluid balance. They also reabsorb essential nutrients and minerals. The loss of 50% of kidney function is like losing one kidney. Healthy people can usually survive on one kidney, but at times of particular stress this might not be adequate.

In another experiment, where mercury fillings were placed in the teeth of monkeys, it was shown that mercury from the 'silver fillings' dramatically altered the normal bacteria of the gut (16). Mercury-resistant strains of bacteria appeared, and it was found that these strains were often resistant to antibiotics as well. These effects of exposure to mercury may help to explain the rise in antibiotic resistant bacteria and the ineffectiveness of antibiotics, which has become a considerable problem in medicine.

Recent reports implicate mercury in certain brain dysfunctions. Autopsy data from patients who died suffering from Alzheimer's disease reveal higher concentrations of mercury in brain areas associated with memory than those in an age-matched group who had not developed Alzheimer's disease (17,18). Others have isolated a biological defect caused by mercury (19) which results in the development of nerve tangles that are characteristic of Alzheimer's disease.

Scientific conclusions

General conclusions:

- Mercury vapour is an extremely dangerous poison.

- There is no absolutely safe level of mercury vapour exposure for humans.

- 'Silver' dental fillings contain 50% mercury.

Conclusions specific to dentistry:

- Mercury is released continuously from 'silver' dental fillings.

- 'Silver' dental fillings release a pharmacologically significant daily dose of mercury in humans – as documented in the work of Dr. Richardson on the risk assessment of amalgam fillings of the Canadian government.

- 'Silver' dental fillings are the largest source of elemental mercury exposure in the general population (as shown in the Environmental Health Criteria 101:Methylmercury, WHO 1990).

- Mercury from 'silver' dental fillings collects in all the adult tissues of the body, being highest in the liver and kidneys.

- Mercury from 'silver' dental fillings crosses the placenta and collects in the developing baby, being highest in the liver and kidneys.

- Mercury from 'silver' dental fillings reduces kidney function.

- Mercury from 'silver' dental fillings alters the normal bacterial population in the gut, contributing to antibiotic resistance.

- Mercury has been implicated in Alzheimer's disease. 'Silver' dental fillings provide a ready source of mercury for brain contamination.

Scientific evidence overwhelmingly confronts the unsupported opinion of the dental profession

The issue of the safety of 'silver fillings' is no longer open to debate. These fillings should not be used.

The symptoms of low level chronic exposure to mercury vapour

Chronic exposure to low level mercury vapour can cause a number of non-specific, seemingly unrelated symptoms leading easily to misdiagnosis. Indeed, this had led to mercury exposure as a cause of ill health being labelled 'the great masquerader'. The similarity of mercury-related symptoms to other medical conditions makes it very difficult for doctors to make a correct diagnosis.

Unfortunately, very few doctors are aware of the fact that mercury is released from dental fillings, and this confounds the possibility of correct diagnoses being made.

Examples of symptoms:

1. Psychological disturbances

- Irritability
- Nervousness
- Shyness or timidity
- Loss of memory
- Lack of attention
- Loss of self-confidence
- Disturbance of intellectual capacity
- Lack of self control
- Fits of anger
- Depression
- Anxiety

- Drowsiness or insomnia

2. Oral symptoms

- Bleeding gums
- Loose teeth
- Bone loss around teeth
- Excessive salivation
- Foul breath
- Metallic taste
- White patches on gums and inside cheeks
- Inflammation of the mouth:
- Ulceration of the gums, palate and tongue
- Burning mouth
- Tissue pigmentation

3. General health symptoms

(a) Gastrointestinal

- Regular abdominal cramps
- Chronic constipation or diarrhoea
- Chronic gastrointestinal problems, including colitis

(b) Cardiovascular

- Irregular heart beat
- Feeble or irregular pulse
- Alterations in blood pressure
- Pain and/or pressure in the chest

(c) Neurological

- Fine tremors (hands, feet, lips, eyelids or tongue)

- Chronic headaches
- Dizziness
- Ringing in the ears
- Numbness of hands or feet

(d) Respiratory

- Persistent cough
- Emphysema
- Shallow or irregular breathing

(e) Immunological

- Allergies
- Asthma
- Rhinitis
- Sinusitis
- Swollen lymph nodes – especially the neck
- Auto-immunity, rheumatoid arthritis, lupus etc

(f) Endocrine (glandular or hormonal)

- Sub-normal body temperature
- Cold clammy skin
- Excessive perspiration

(g) Skin

- Sores on the skin – especially face and shoulders

(h) Kidney

- Electrolyte imbalance – i.e. mineral imbalance
- Reduction in kidney function

- Renal failure in severe cases
- Glomerular nephritis

(i) Other

- Muscle weakness
- Chronic fatigue
- Joint pain
- Anaemia
- Loss of appetite
- Constriction of visual field – tunnel vision

The symptoms of mercury exposure are very varied and confusing as they cover such a wide and diverse picture, and they can mimic many other diseases.

How mercury from dental fillings causes disease

Mercury is a toxic heavy metal, which even in very small quantities can be very detrimental to human health. It affects the body in 4 main ways:

1. **Poisoning**
2. **Allergic reactions**
3. **Induction of auto-immunity**
4. **Idiosyncratic reactions**

1. Poisoning

Mercury is a heavy metal, which, like lead and arsenic, is very toxic to the human body.

Mercury is liquid at room temperature and so vapourises very easily. The vapour is odourless, colourless, tasteless, and when inhaled is very rapidly taken up by the blood and passes to all the other tissues of the body.

Once in the tissues it can convert to inorganic mercury compounds. These highly reactive chemicals react with the sulphur-containing enzymes. Enzymes are special substances that enable essential chemical reactions within the cells that allow the cells, and therefore the tissues and body, to function. Without enzymes we would not be alive. More than 99% of the enzymes in the body contain sulphur. Consequently the main mode of action of mercury in the cells is as an enzyme poison, with the final result being metabolic disease.

2. Allergy

Allergic reactions to mercury and other constituents of 'silver fillings' can occur either locally around the teeth and gums, or in the oral tissues. These reactions may occur immediately, or they may take up to several years to develop. Symptoms may appear in other parts of the body (systemic), and these also may appear immediately, or take several years to appear. Eczema and sores on the face or shoulders are typical systemic reactions.

Conventional dental opinion is that less than 1% of the population is sensitive to mercury. However, it is scientific fact that approximately 5% of the population is allergic to mercury. This means that 5 out of every 100 patients in a dental practice may be allergic to it.

3. Mercury-induced auto-immunity

The immune system is the body's defence system. It protects 'you' from things that are 'not you' – so it protects against such things as infections and cancer. Autoimmunity is a reaction in the immune system when that system can no longer tell the difference between 'you' and 'not you', and so starts to attack healthy tissues. There are many specific auto-immune diseases – e.g. multiple sclerosis, lupus and rheumatoid arthritis.

If you have any of these diseases, and are told that having your 'silver fillings' removed will cure you, please be very cautious. Choosing to have the mercury removed from your mouth is a very positive move for your future health. Although this will reduce your daily exposure, the removal of 'silver fillings' is not a scientifically proven cure for disease.

4. Idiosyncratic reactions

Some people may be immunologically hypersensitive or toxicologically sensitive to very low doses of mercury. Some individuals have idiosyncratic (unusual) reactions to mercury and its compounds. Often the symptoms are bizarre and can be quite difficult to diagnose. The symptoms may be included in a whole range of environmentally-induced reactions that are unique to any particular individual.

Conclusion

One of the important questions facing medical research today is whether mercury from 'silver fillings' causes specific diseases. Many diseases today have no known cause – and none of these have been proved to be caused by mercury from fillings. However, mercury is even more poisonous than either lead or arsenic, and exposure to it, no matter how small, will cause some level of mercury poisoning. The negative effect mercury will have on health should never be ignored. Removal of mercury using safe procedures will be a positive health move.

Showing the impact that low level exposure to mercury vapour has on your health is very difficult. There is not one single test that can show this, nor any tests that are 100% accurate.

Lab tests will help to evaluate the following:

- Overall health status
- Environmental & life style factors
- Co-existing medical factors
- Need for therapeutic nutrition

- Liver & kidney function during removal of fillings that contain mercury
- Excretion from the body during removal of fillings that contain mercury

When should fillings containing mercury be removed?

In our own dental practice we do not use 'silver fillings'. If a 'silver filling' needs to be removed, we do not replace it with another 'silver filling'. Where possible, such a filling is replaced with a metal-free restoration – preferably what we call a 'composite'. The filling options will be outlined to you later in this booklet.

Other times when 'silver fillings' should be removed are outlined below.

- At your own request – having decided to make a positive health choice. After all, it's your body, and you have the choice of what you fill your teeth with.

- At your doctor's recommendation – i.e. if in the doctor's opinion your general health is being affected by mercury and that removal of your fillings will aid your recovery.

- If you have a positive patch test for mercury or have other proof of an allergic reaction to the mercury in your fillings. At the moment approximately 5% of people are allergic to mercury.

- If your dentist feels that the mercury is having a detrimental effect on your teeth, gums and/or oral tissues.

The above information represents reasonable criteria for the removal of 'silver fillings' based on current published scientific data and clinical experience.

What are the alternatives to 'silver fillings'?

'Silver fillings' have been the material of choice for the last 150 years. Because they are relatively cheap and simple to use, consideration of their biocompatibility was of little or no interest in those days, and its significance was overlooked.

The alternatives tend to be a little more expensive and clinically more demanding in the application. They fall into 3 main categories:

- **Gold alloy restorations**
- **Porcelain restorations**
- **Composite restorations**

Gold alloys are a mixture of gold and other metals to make the gold harder. They are usually cast to make inlays, onlays or crowns.

The advantages of gold restorations are:

- They can be used on the biting surface of teeth.
- They do not corrode very readily.
- They fit very accurately.
- They are made in the lab and the bite is made to fit correctly.
- They fit between the adjacent teeth so food will not get stuck between your teeth.

The disadvantages of gold restorations are:

- They require a high level of skill to prepare and fit.
- They are held in place with dental cement, and this does not strengthen the tooth.
- They are not tooth-coloured, and some people find them unsightly.
- Some people are sensitive to the metals in the alloys.

Porcelain is made by fusing minerals such as feldspar silica and alumina in a glass matrix at high temperatures to form a translucent material that is very tooth-like in appearance.

The advantages of porcelain restorations are:

- They have a natural tooth-like appearance.
- They are very durable and collect very little plaque.

The disadvantages of porcelain restorations are:

- They are very brittle.
- They tend to wear the opposing teeth excessively
- It is very difficult to get an accurate fit.

Composite fillings are made from a mixture or finely ground quartz and Bis-GMA resin. They have very tooth-like qualities in both appearance and hardness. They were originally developed for anterior teeth because of their tooth-like appearance, but more recently the amount of quartz filler has been increased and the particle size decreased to give them increased wear resistance. The composite restorations may be fabricated either directly in the mouth (<u>direct</u>), or made in the lab (<u>indirect</u>).

<u>Direct composite fillings</u> are mixed by the dentist at the time of use.

The advantages of direct composite fillings are:

- Bio-compatibility
- They are tooth-like in appearance and strength.
- The possibility of electrical currents in the mouth is removed (providing no other metal is present in the mouth).
- They enhance overall tooth strength due to bonding of the fillings to the tooth.
- They are cost effective.

The disadvantages of direct composite fillings are:

- 70% of the material polymerises in the mouth, the remaining material does not react and so stops the filling from being quite as strong as it might be.
- Some people have an allergic reaction to the un-polymerised material, especially those sensitive to petroleum products.
- The composite shrinks slightly on setting and so if the bond is not good it may leave an open margin which may lead to sensitivity or the filling not to last as long as we might hope.
- Composites are very demanding of the dentist's skill.

<u>Indirect composite fillings</u> are processed in the lab using heat, light and pressure. This allows more of the material to be polymerised, and so the shrinkage is controlled.

The advantages of indirect composite fillings are:

- The lab controls the shrinkage so that when the filling is bonded there are no open margins.
- The lab-processed fillings look like teeth, wear like teeth, and are able to be made with great precision. They are the most compatible tooth restoration.
- They have all the advantages of the gold and porcelain restorations with none of the disadvantages.
- They are a cost-effective durable restoration.

The disadvantages of indirect composite fillings are:

- Cost
- More time-consuming to produce

Are these alternative materials safe?

The ability to measure toxic materials has always been difficult, and their effect on the body is almost impossible to determine, due to the individuality of each person.

However, when Dr. Richardson carried out the equivalent tests on composites as he did on amalgam for the Canada Health 2000 Group (22,23) he found no significant risk in the use of composites. In fact, he announced that the likelihood of oestrogenic effects from bisphenol-A in patients with 8 composites was 125 times lower that the level considered to be safe, whereas patients with 8 amalgam fillings are exposed to quantities of mercury that are 3 times higher than levels that are considered to be safe.

Antibodies are usually produced when toxic elements are introduced to the body. One way of assessing the toxic effect of a material is to measure these antibodies as carried out in the Clifford Materials Reactivity Test. This test is useful when treating environmentally compromised or highly allergic patients.

Applied kinesiology is a well-researched technique that is used in some holistic dental practices to help to assess the biocompatibility of the various materials. It involves testing the body's reaction to a material by using the strength of various muscles as the indicator.

What precautions should be taken during the removal of 'silver' fillings?

While the fillings are being drilled out, heat and small particles of filling are produced. Consequently there is a high risk of mercury exposure at these times. This risk can be minimised by taking a number of important precautions.

- The use of rubber dam, which is a thin sheet of rubber that isolates the teeth from the rest of the mouth, and protects the patient from swallowing particles of filling. It also reduces the amount of vapour inhaled.

- The use of special suction devices that cover each individual tooth as it is being drilled. The use of high volume vacuum suction.

- The use of an alternative source of air to breathe. This reduces the chance of inhaling mercury vapour that is released as a result of the drilling.

- Using room fans and ionizers to keep the air moving and carry the mercury vapour away from the patient.

- The use of protective covering and eye protection for the patient, as the skin absorbs mercury on contact. The eyes are especially vulnerable.

- The use of copious amounts of cold water when drilling. This reduces the amount of vapour produced, as well as reducing the trauma to the tooth from the heat produced while drilling.   In our practice we also use a chelating agent to bind with the vapour and make the mercury less available to the body.

- By sectioning the filling instead of grinding it out the amount of vapour is reduced.

- Homeopathic remedies may be offered, as well as nutritional support and activated charcoal. Such remedies help the body to excrete any mercury that might have been absorbed during the removal process.

Detoxification

A patient may become aware of suffering symptoms that are likely to be the result of long term exposure to mercury from 'silver' fillings, and may wish to find out what to do to improve health.

There are safe, effective ways of removing mercury and mercury compounds from the body. Such methods must always be supervised by a competent practitioner.

Examples of these methods are:

Nutritional therapy

To minimise the effect of the mercury, specific nutritional supplements can be taken.

Certain minerals and vitamins affect the way in which the mercury is metabolised and taken up by the body. Simply put, these minerals and vitamins are either used to make new enzymes to replace the ones destroyed by the mercury, or they attach to the mercury and make it less bio-available. This is a very complicated subject – if you wish to know more about it, please ask, and we can give you more information.

Chelation therapy

This employs the use of drugs that have a strong affinity for mercury in order to eliminate it from the body. These drugs are not without side-effects, and should only be given under the supervision of a doctor. They are very efficient at reducing the body's mercury burden.

Field Control Therapy (FCT©)

It is important to state that other therapies cannot be used in conjunction with this approach.

FCT is a more recently-devised method, and it has been found to be more efficient than other therapies in its capacity to help the body to excrete mercury.

This system involves the use of tiny amounts of specifically energized water taken under the tongue to provoke the tissues and organs of the body to release mercury and mercury compounds that are stored there. This release must take place in a carefully controlled environment to ensure that the mercury that has been mobilised leaves the body. This requires the patient to stay away from sources of strong electro-magnetic fields for a set period of time for each treatment session. Patients are also required to follow a daily diet that excludes all foods that might interfere with the treatment. This diet should be adhered to throughout the overall treatment period, not just when remedies are being taken.

The practitioner supervises the process carefully by testing the patient's system regularly (approximately monthly at first) and keeping a meticulous record of the patient's reported symptoms and reactions.

It is recommended that the patient undergoes a treatment regime every few weeks until health improves. After satisfactory improvement, it is advised that the patient is re-tested approximately every three months.

Bibliography

1. Phillips, R. W. (1982): Skinners science of dental materials 8th Ed. Philadelphia: W. B. Saunders Co. p311
2. Paterson, N. (1994): The longevity of dental restorations. Br. Dent. J. 157: 23-25
3. Vimy, M.J. & Lorscheider, F.L. (1985): Intra-oral air mercury released from dental amalgam. J. Dent. Res. 64: 1069-71
4. Vimy, M.J. & Lorscheider, F.L. (1985): Serial measurements of intra-oral air mercury: Estimation of daily dose from amalgam. J. Dent. Res. 64: 1072-5
5. Patterson, J.E., Wiessberg, B.G. & Dennison, P.J. (1985): Mercury in human breath from dental amalgam. Bull. Environ. Contam. Toxicol. 34: 111-123
6. Vimy, M.J. & Lorscheider, F.L. (1990): Dental amalgam mercury daily dose estimated from intra-oral vapour measurements: A predictor of mercury accumulation in human tissue. J. Exper. Med. 3: 111-123
7. Fredin, B. (1988): Studies on the mercury release from dental amalgam fillings. Swed. Dent. J. 3: 8-15
8. Nylander, M., Friberg, L. & Lind, B. (1987): Mercury concentrations in the human brain and kidneys in relation to exposure from dental amalgam fillings. Swed. Dent. J. 11: 179-187
9. Eggelston, D.W. & Nylander, M. (1987): Correlation of dental amalgam with mercury in brain tissue. J. Prosth. Dent. 58: 704-707
10. Clarkson, T.W., Friberg, L., Hursh, J.B. & Nylander, M. (1988): The prediction of intake of mercury vapour from amalgams: In Biological Monitoring of Toxic Metals

pp.247-260 Plenum Press, New York

11. Truono, E.J. (1991): Letter of importance. JADA 122: 8-14

12. ADA. When your patients ask about mercury in amalgam. JADA 120:395-8. April 1990

13. Denscher, G., Horsted-Bindslev, P. & Rungby, J. (1990): Traces of mercury in organs of primates with amalgam fillings. Exp. Mol. Path. 52: 291-299

14. Vimy, M.J., Takahashi, Y. & Lorscheider, F.L. (1990): Maternal-fetal distribution of mercury (203-Hg) released from dental amalgam fillings. Am. J. Physiol. 258: R939-R945

15. Vimy, M.J., Boyd, N.D. & Lorscheider, F.L. (1990): Glomerular filtration impairment by mercury released from dental 'silver' fillings in sheep. Am. Physiol. Soc. Fall meeting Orlando Fl. Oct. 9. 1990 The Physiologist 33(4), 94, 1990

16. Summers, A.O., Wireman, J., Vimy, M.J., Lorscheider, F.L., Marchall, B., Levy, S.B., Bennet, S. & Billard, L. (1993): Mercury released from dental 'silver' fillings provokes an increase in antibiotic resistant bacteria in primate oral and intestinal flora. Antimicrobial Agents & Chemotherapy 37: 825-834

17. Thompson, C.M., Markesbery, W.R., Ehmann, W.D., Mao, Y-x. & Vance, D.E. (1988): Regional brain trace-element studies in Alzheimer's disease. Neurotoxicol. 9: 1-7

18. Wenstrup, D., Ehmann, W.D. & Markesbery, W.R. (1990): Trace element imbalances in isolated subcellular fractions of Alzheimer's disease brains. Brain Res. 533: 125-131

19. Duhr, E., Pendergrass, C., Kasarskis, E., Slevin, J. & Haley, B. (1991): Hg2+ induces GTP-tubulin interactions in rat brain similar to those observed in Alzheimer's disease. FASEB J. 5: A456

20. Richardson, G.M. & Allan, A.M. (1996): Monte Carlo Assessment of mercury exposure and risks from mercury amalgam. Human Ecol. Risk Assess. 2(4): 709-761
Richardson, G.M.: Environmental Health Directorate, Health Canada: Assessment of Mercury Exposure and Risks from Dental Amalgam, 1995, Final Report.
21. Boyd, N.D., Benediktsson, H., Vimy, M.J., Hooper, D.E. & Lorscheider, F.L. (1991): Am. J. Physiol. 261 (Regulatory Integrative Comp. Physiol.30): R1010-R1014
22. Richardson, G.M. (1997): An assessment of adult exposure and risks from components and degradation products of composite resin dental materials. Human Ecolog. Risk Assessment. 3(4): 683-697
23. Richardson, G.M. and others (1999): Preliminary estimates of adult exposure to Biosphenol-a from dental materials, food and ambient air. Environmental Toxicology and Risk Assessment: 8th Volume, ASTM STP 1364. Henshel, D.S. and others: Eds, ASTM, West Conshocken, PA

Other sources of information

British Dental Association (BDA) – amalgam fact file
www.bda.org/dentists/policy-research/bda-policies/public-health/fact-files/amalgam.aspx
This fact file discusses the subject of dental amalgam in general, covering use and health effects. It also details UK government and international perspectives.

British Dental Health Foundation
www.dentalhealth.org.uk

USA Government Department of Health and Human Services. Centers for Disease Control and Prevention: Division of Oral Health.
Dental amalgam – Use and Benefits – Fact Sheets and FAQs…
www.cdc.gov/oralHealth/publications/factsheets/amalgam.htm

International Academy of Oral Medicine and Toxicology
www.iaomt.org

Dental Amalgam Mercury Solutions (DAMS)
www.amalgam.org   e-mail: dams@usfamily.net
Telephone helpline: +1 651 644 4572
Phone or e-mail for current information on dental amalgam problems. Information is also available on a wide range of other dental issues.

List of relevant books

Dentistry without Mercury (1995) by Sam Ziff and Michael Ziff D.D.S   Bio-Probe, Inc. ISBN 0-941011-04-6

Silver Dental Fillings: The Toxic Time Bomb (1994) by Sam Ziff (1994) Thorsons ISBN 0-7225-1232-5

Uniformed Consent: Hidden Dangers in Dental Care (1999) by Hal A. Huggins D.D.S., M.S. and Thomas Levy M.D., J.D. Hampton Roads Publishing Company, Inc.
ISBN 1-57174-117-8

Toxicology of Metals (1996) edited by Louis W. Chang, L. Magos and Tsuguyoshi Suzuki   Lewis Publications or CRC Press ISBN 1-56670-803-6

See also:

A Layman's Guide to Field Control Therapy

by

Stewart J Wright BDS

www.ingramcontent.com/pod-product-compliance
Lightning Source LLC
Chambersburg PA
CBHW071335200326
41520CB00013B/2996